YOUR KNOWLEDGE HAS VALUE

- We will publish your bachelor's and
  master's thesis, essays and papers

- Your own eBook and book -
  sold worldwide in all relevant shops

- Earn money with each sale

Upload your text at www.GRIN.com
and publish for free

Miriam Herbert

Aus der Reihe: e-fellows.net stipendiaten-wissen

e-fellows.net (Hrsg.)

Band 1096

Boon and Bane of not Being Subject to the Hayflick Limit

What effects does it have in cancer cells and what is it used for in life extension science?

GRIN Publishing

Bibliographic information published by the German National Library:

The German National Library lists this publication in the National Bibliography; detailed bibliographic data are available on the Internet at http://dnb.dnb.de .

Imprint:

Copyright © 2014 GRIN Verlag GmbH
Print and binding: Books on Demand GmbH, Norderstedt Germany
ISBN: 978-3-656-88239-8

GRIN - Your knowledge has value

Since its foundation in 1998, GRIN has specialized in publishing academic texts by students, college teachers and other academics as e-book and printed book. The website www.grin.com is an ideal platform for presenting term papers, final papers, scientific essays, dissertations and specialist books.

Visit us on the internet:

http://www.grin.com/

http://www.facebook.com/grincom

http://www.twitter.com/grin_com

Boon and bane of not being subject to the Hayflick limit

-

What effects does it have in cancer cells and what is it used for in life extension science?

Miriam Herbert

5. Prüfungskomponente

Abitur 2014

Index

I. Introduction

All living things have to die. This fundamental truth is held to apply even to the smallest unit of life – cells. However, there is a phenomenon that is sometimes called biological immortality. It refers to cells that live beyond their proclaimed life span, which is roughly set by the Hayflick limit. All cancer cells have acquired this property; they divide indefinitely, which is the essential problem with cancer cells. On the other hand, researchers are very much interested in the molecular mechanism behind this property to may be able to use it to extend life and rejuvenate cells. Cells that are not subject to the Hayflick limit are generally seen as a threat to the human body, however, they are interesting subjects of experiments and scientists have already learned a great deal of knowledge by studying these mutants and continue to gain more important insights into the functioning of any kind of human body cell. Immortal cells can be boon and bane for humankind. Certain aspects of this issue will be discussed.

II. Hayflick Limit

i. History of the Hayflick Limit

As mentioned above, normal cells are limited in their capacity to divide. This limit is called Hayflick limit, a term coined in 1974 by Sir Macfarlane Burnett[1], for Leonard Hayflick was the one who demonstrated that cells are not able to divide indefinitely.

The idea was not new at the time; in 1881, the German biologist August Weismann proposed:

> "Death takes place because a worn-out tissue cannot for ever renew itself, and because a capacity for increase by means of cell-division is not everlasting, but finite. [...] Functional disturbances will appear soon as the rate at which the worn-out cells are renewed becomes slow and insufficient."[2]

Weismann could not prove his thesis though and it was soon to be almost entirely forgotten. By the time Hayflick performed research on cell culturing, Weismann's concept had been overturned by Alexis Carrel's hypothesis that every human cell in culture can proliferate indefinitely if provided with the correct medium and nutrition. He based this proposition on his experiments with chicken heart fibroblasts he supposedly grew in culture for more than 20 years[3], which is far more than a normal chicken's lifespan.

In 1958, Leonard Hayflick commenced research into the possible viral etiology of cancer. However, he soon came across certain difficulties. Initially, he intended to expose normal human body cells to cancer cell extracts, to hopefully observe cancer-like changes in the normal cells, but the normal cells died soon. According to the generally accepted thesis of Carrel, Hayflick sought the problem in his experimental procedure and set-up. He started a carefully conducted series of experiments over three years, which eventually convinced him that his experiments did not fail due to a technical mistake.[1] In 1961, Hayflick and the cytogeneticist Paul Moorehead set up further experiments. They mixed equal amounts of normal human male fibroblasts at about their fortieth population doubling with normal human female fibroblasts at about their tenth population doubling. Unmixed controls of the two populations were kept. When the "old", male control group stopped dividing, the mixed culture was examined. In fact, there were only female fibroblasts remained. This proved that the "old" cells stopped dividing even in the presence of "younger" cells, which continued to replicate and therefore a technical error or viral contamination were rather improbable explanations as to why only the male cells stopped dividing and died.[4]

Other experiments showed that cells even "remembered" their age, after they had been frozen for some time, and stopped dividing after around 50 population doublings (human fibroblasts)[5] regardless of the age they were at when they were frozen.[1] The numbers differ greatly between species and individuals; most importantly, these observations indicate that there is some molecular mechanism which "counts" the divisions a cell completed and stops the replicative process as soon as the Hayflick limit is reached.[5] This leads to another question: how does the Hayflick limit function on the molecular level?

ii. The End Replication Problem

Key to this question is the characteristic of the Hayflick limit to be determined by the number of divisions, not by time. Therefore the answer must be connected to the replication process of a cell, mitosis. The whole genome of the cell must be replicated before mitosis, so each new daughter cell can have the exact copy of genetic material.[6] DNA is replicated in a semiconservative manner. Each of the two DNA-strands functions as a template for a new complementary strand, which leaves the cell with two exact copies of the same double-stranded DNA.[7]

DNA replication is a complex multistep process. Initially, initiation proteins locate particular points of the DNA known as origins of replication. An enzyme called helicase unwinds and splits the two strands at these sites and replication forks form in both directions. The forks consist each of a leading strand and a lagging strand. The leading strand is synthesized continuously. The enzyme primase assembles a short RNA primer to which a DNA polymerase attaches new nucleotides in 5'→3' direction, whereat the existing DNA strand serves as a template for the new, complementary strand. Another enzyme named RNase H removes the RNA primers and DNA polymerase fills in the gap. Finally, all DNA fragments are joined together by the enzyme ligase to form a stable new strand.[7]

The process of synthesizing a complementary strand to the lagging strand is a little more complicated. DNA polymerase can add nucleotides to a nucleic acid only to its 3'-end. Therefore the lagging strand must be synthesized discontinuously. Primase has to synthesize primers within short distances. The DNA polymerase starts adding nucleotides to the first primer until it reaches the next primer. Doing so, it creates fragments of complementary DNA, called Okazaki fragments, between two primers. Afterwards the enzyme skips ahead to the next primer upstream and assembles a new Okazaki fragment. The lagging strand is synthesized fragment by fragment and again RNase H degrades the primers, DNA polymerase fills in the gaps and ligase joins all fragments together. This process works flawlessly with circular DNA;[8] however, eukaryotic cells have linear chromosomes. This offers an advantage in recombination and enhances genetic diversity, but it also poses a new problem to the cell. Since DNA polymerase can add nucleotides only to existing nucleic acids and only in 5'→3' direction, it cannot replicate the very ends of linear chromosomes. Even if the last primer would line up perfectly with the chromosome end, the primer would eventually be degraded and, again, leave a single-stranded 3'-end exposed. This problem is termed the end replication problem.[9]

This is an issue concerning chromosome stability and DNA erosion.[44] The terminal overhang resembles a double-strand break and could be recognized by the cell's repair mechanisms as such, which could lead to chromosome aberration and eventually to apoptosis.[8] The unstable ends might fuse with other ends, amid another chromosome or even with the nuclear membrane and form miscellaneous, unfavorable chromosome forms. Besides, the end replication problem causes the DNA to shorten with every replication cycle, since not all of it can be replicated. This erosion might lead to the loss of gene function or mutations.[10]

To prevent these problems, linear chromosomes have repetitive non-coding sequences at every end. In human beings, the sequence is 5'-TTAGGG-3'.[7] The existence of such non-coding sequences at the tips of chromosomes had been known since 1938; however no one could plausibly explain what they were good for. One of the involved geneticists, Hermann J. Muller, coined the term telomeres for these terminal regions from the Greek *telos* "end" and *menos* "part". These "end parts" have varying lengths in different species and individuals.[1] Human beings' telomeres are on average about 5,000-15,000 bp long and are shortened by approximately 100 bp with every cell cycle.[5] These non-coding regions can safely be shortened during every cell replication without functional losses of the genome. Also, it has been shown that telomeres are different from normal linear nucleic acid ends. Most of the telomere region is double-stranded, but it ends in a long 3' one-stranded overhang, which forms a loop resembling a knot, which is called t-loop. This part may also develop four-stranded formations called G quadruplexes. They make telomeres distinguishable from double-strand breaks.[8]

The functions telomeres fulfill are vital to the cell and repetitive sequences of DNA alone are not enough to maintain chromosomal stability. In fact, telomeres are nucleoprotein complexes consisting of the mentioned repetitive sequence and associated proteins. The telomeric protein complex and its associating molecules have not yet been fully deciphered.[9] In humans, a few proteins like TRF1/TRF2, POT1/POT2, and TIN1/TIN2 are known to be sequence specific proteins binding to 5'-TTAGGG-3'. They are referred to as shelterin complex, because their functions include protecting the protrusions from fusion, chemical modifications and nucleases, holding the t-loops and quadruplexes together, as well as regulating telomere length.[11]

Even if telomeres fulfill their tasks perfectly, the cell will at some point reach senescence, because telomeres are not a remedy for the end replication problem, but only a deferment.[1] When the telomeres are too short to sustain chromosomal stability, a cryptic splice site is activated leading to a mutant form of the lamin A protein called progerin.[12]

This means that the RNA transcript for the protein is spliced in an alternative way, which is also actively involved in the premature aging disease Hutchinson-Gilford progeria syndrome (HGPS). Lamin A is a structural protein in the nucleus. Progerin is a pared version of lamin A and thus cannot fulfill its tasks properly, which leads to a deformed nucleus. Short telomeres are thought to yield further alternative splicing in multiple other genes collectively causing the cell to stop dividing.[12] This state is referred to as senescence and leads eventually to cell death. However, there remains a slight chance for the cell to overcome this crisis, if it manages to elongate its telomeres and overcome the Hayflick limit.[13]

iii. Telomerase

Telomeres are found in every human body cell. However, they are longest in germ line cells.[7] This is necessary, because human beings develop from one single zygote, which is subject to numerous cell divisions to form a new body. This would not be possible, if its telomeres shortened with every cell cycle and the cell soon reached the Hayflick limit. The natural remedy to this problem was discovered by Carol W. Greider, Elizabeth H. Blackburn and Jack W. Szostak, who received the 2009 Nobel Prize in Physiology or Medicine for their findings. They proved the existence of an enzyme called telomerase.[14] It is a ribonucleoprotein reverse transcriptase, which is able to elongate telomeres.[8]

Telomerase consists of a RNA subunit and a protein subunit. The protein compound called TERT surrounds the RNA subunit called TR.[15] TR is 11 bp long and complementary to the telomeric sequence: 3'-CAAUCCCAAUC-5'. It serves as a template to add new repeats to the existing telomeric DNA strand.[8] The single-stranded end of a telomere ends in 5'-TTAGGGTTAG-3'. hTR (human telomerase RNA) bonds to the existing DNA beginning from the last G of the last complete TTAGGG sequence and then adds six new nucleotides, 5'-GGTTAG-3', by serving as a template for DNA-polymerase. TERT acts as a catalyst in this reaction and shields it from chemical disturbances.[15]

Telomere lengthening through telomerase is highly regulated.[16] First, the constituting components have to be assembled and brought together. The latter happens in Cajal bodies, sub-organelles, which shelter and transport telomerase to telomeres. These telomerase-containing Cajal bodies can be recruited to the telomere sites during S phase of the cell cycle, when the whole genome is replicated and telomeres have to be uncapped.[17] The process of how telomeres and the telomerase in the Cajal bodies are brought together is incompletely understood. Recent studies suggest a participation of the Homeobox binding protein 1 (HOT1). HOT1 binds to double-stranded telomere regions as well as to the outer layer of Cajal bodies and brings these two into close proximity. Furthermore, experiments with mice showed that HOT1 is most likely necessary for Cajal bodies to bind to chromatin.[17] Still, this model needs further testing and refinement to fully understand this complex regulatory pathway.

Telomerase is not only expressed in germ line cells, but also in tissues with high turnover.[15] These types of cells are referred to as telomerase positive cells and include white blood cells, testis tissue, intestinal villi, bone marrow, skin cells, pulmonary tissue and lymph nodes.[8] The activity of telomerase in these cells is considerably lower than in germ line cells, thus it is not sufficient to elongate telomeres, but it slows the degradation.[16] Obviously, telomerase is crucial to life, but reactivation of telomerase in somatic cells could enable them to circumvent the Hayflick limit with disastrous consequences.

III Cancer Cells

i. Telomeres in Age-Related Diseases

Just as Weismann suggested in 1881, age-related diseases suggest themselves to be associated with shortened telomeres, which cause cells to become senescent so "worn-out tissue cannot [...] renew itself".[2] Prof. Carol Greider et al. performed further experiments with mice to show the correlation of age-related diseases and the absence of telomerase which furthermore implies the correlation with shortened telomeres, too.[18] They knocked out the genes for telomerase in a generation of mice. They called this generation G1 and continued to breed the mice for a couple of generations. Analyses made evident that short telomeres accumulated continuously with progressive generations. Also, the telomeres seemed to lose their functions and chromosome fusions became common. Strikingly, cellular senescence and apoptosis became only from the G4 generation on a wide spread phenomenon that lead to tissue deformation. Therefore it must truly be the shortened telomeres, not the absence of telomerase in general that lead to malformation.[15] The hypothesis is that too short telomeres lose their protection and appear like a double-strand break to the cell, which triggers a signal transduction pathway leading to either senescence or cell death. The genotypic observations were reflected in the phenotypes, too.[19] The knock-out mice were more prone to age-related degenerative diseases in general, more often infertile, showed reduced cellularity in e.g. bone marrow and spleen tissue, had less B and T lymphocytes, and experienced high rates of cancer.[15] The later the generation, the shorter was the life expectancy. Prof. Greider and Cal Harley also analyzed the telomeres of white blood cells from people of different age and found a progressive decline of telomere length with years.[20] This confirms that even telomerase positive cells can show telomere shortening, if the loss of telomeric DNA is greater than the expression of telomerase.

Extremely short telomeres may also lead to the loss of tumor suppressor mechanisms, because they forfeit their protecting and stabilizing functions.[15] This in effect promotes disorder in the cell[45] and abets the worst case scenario: cancer.

ii. Telomerase in Cancer

Cancer is a group of diseases marked by uncontrollable cell growth, most of the time in forms of malignant tumors.[10] In 2000, two cancer researchers, Douglas Hanahan and Robert Weinberg, attempted to define distinct capabilities of cancerous cells. They found six hallmarks of malignant tumors.

In order to become a malignant cancer, a cell has to be able to:

1. divide independently from external growth factors
2. ignore external tumor suppressor signals
3. avoid apoptosis
4. divide indefinitely without senescence
5. stimulate stained angiogenesis
6. invade tissue and establish distant secondary tumors

Even though the last hallmark is under constant debate, several mutations must occur within the concerned cell's genome to acquire these properties.[9] According to point three and four, immortalization is necessary for tumorigenesis. The Hayflick limit is therefore a major tumor suppressor mechanism. It prevents unlimited cell growth and protects genes from mutation through erosion.

Unfortunately, this mechanism is not perfect. There are cells which overcome the Hayflick limit and undergo carcinogenesis. Several steps can lead them there.[20] The terminal proliferation arrest (TPA) of the Hayflick limit can be extended through the mutation of the p53 gene. If a senescent cell loses function of p53, it can divide a little longer, but soon reaches the next TPA referred to as p53-minus TPA. The loss of expression of another gene called pRb also allows for a further finite extension of the cell's proliferative lifespan. Both mutations can be combined. However, the cell then enters a state called crisis. Crisis is the ultimate barrier to cell immortalization.[21] Cells that escape crisis must all activate some kind of telomere maintenance mechanism. In the vast majority of cancer cells, this happens to be the reactivation of telomerase. Telomerase is present in about 90% of all human cancers.[8]

Studies have shown that the in vitro introduction of telomerase cDNA to senescent cells is enough to enable them to overcome senescence and crisis.[13] On the other hand, it has not been clarified whether the activation of telomerase is usually a cause or a product of mutations in oncogenes. The few human papilloma virus (HPV) strains that are associated with forming malignant tumors are the same that are also able to immortalize normal human cells in vitro. This indicates that immortalization makes cancer feasible.[20] On the other hand, it is hard to obtain viable, immortal cell lines from many cancers, which would nourish the position that immortalization is a byproduct of tumorigenesis, but not crucial. However, this inability to obtain immortal cell strains from some cancer types might be due to suboptimal cell culturing systems for these specific varieties of cancer.[42] Another argument against immortalization causing cancer is a calculation saying that a cell could create a one kilogram heavy tumor after just 40 population doublings. This number of population doublings would be possible within the Hayflick limit and therefore no immortalization necessary.[21] Then again, this would mean that the starting cell must have had already acquired all mutations needed to proliferate uncontrollably and that none of its daughter cells dies. Both assumptions are rather improbable.[19] Even though there is lots of evidence for immortalization being a prerequisite for cancer, it has not been proven unerringly, too.

Since there is telomerase in most cancers, though, the thought comes to mind that this enzyme might be used as a tumor biomarker in cancer diagnosis. In fact, modern assays like the telomeric repeat amplification protocol (TRAP) are quite efficient and accurate enough to require just a few milligrams of tissue or milliliter of blood to produce a significant result.[22] However, they are so good, they even detect the low quantities of telomerase present in some normal somatic tissues that are telomerase positive, e.g. lymph nodes. Since there is no difference known yet between telomerase in normal calls and cancerous cells, telomerase activity can not be differentiated. Unless scientists learn to differ between "good" and "bad" telomerase expression, testing for telomerase will remain trivial.[23] Even an assay that could make such a distinction would not make a definite diagnosis possible as explained in the following.

iii. Alternative Lengthening of Telomeres

Telomerase activity explains the overcoming of the Hayflick limit in about 90% of human cancers, but what about the other 10%? Hanahan and Weinberg say that immortality is an essential characteristic of cancer. Thus, there is just one reasonable solution: there must be a way of alternative lengthening of telomeres.

First, it had to be proven that there is actually no telomerase involved. Kim et al. examined six immortalized cell lines with the Telomere Repeat Amplification Protocol (TRAP) assay.[20] The assay consists of two steps. First, enzyme-containing cell extracts are mixed with an oligonucleotide primer. Telomerase extends this primer if present. The primer products are amplified by polymerase chain reaction (PCR) and are then feasible to detect.[23,43] The conducting researchers detected telomerase in four of the examined immortal cell lines, but two of them lacked telomerase. KB319, one of the two telomerase negative cell lines, was used in another study. One telomere was observed closely and found to shorten gradually as expected, sometimes the telomere lengthened rapidly. Also, 15 more telomerase negative immortal cell lines were examined which had been artificially immortalized by treatments with carcinogens. The pre-immortalized cells of the same lines were still available and compared to the immortal cells. The comparison showed that telomere lengthening occurred after immortalization. [20]

To exclude the possibility of telomerase inhibitors or the presence of an RNase that might block or degrade telomerase, further experiments were conducted. Lysates from the telomerase negative immortal cell lines were mixed with HeLa cells, a telomerase positive immortal cell line, and tested with the TRAP assay. All probes showed telomerase activity, thus there could not be any telomerase inhibitor or RNase present in the telomerase negative cell lines. This proves that there must be one or several mechanisms of alternative lengthening of telomeres (ALT).[24]

There are four major models of ALT. The first model suggests that the single-stranded telomeric overhang of one chromosome serves as a template for the elongation of another chromosome's telomere by invading the other telomere. The second model assumes that a telomere might be able to serve itself as a template. The t-loop forming telomere might be able to invade its own DNA and synthesize new repeats. However, rapid lengthening of telomeres in telomerase negative immortal cell lines has been observed. This suggests an elongation by means of recombination. The last two models both support the idea of extrachromosomal DNA providing telomeric sequences either in linear or circular form, which recombine with the existing telomeres.[20]

Certain features of ALT expressing cells support the idea of more than one possible ALT mechanism. Telomere lengths in those cells are highly heterogeneous ranging from less than 3000 bp to more than 50000 bp. Additionally, they contain extrachromosomal DNA molecules including telomeric repeats that possibly disintegrated in the process of t-loop formation. These observations would favor a recombination model. On the other hand, the crossing of two telomeres as suggested in the first model has been shown to occur in ALT expressing cells.[20] Another component that might have a role in ALT is a number of protein structures called ALT-Associated PML Bodies (APBs). PML bodies are hollow protein structures associated with surrounding chromatin. Usually they do not contain nucleic acids, but APBs are special PML bodies; they have been demonstrated to contain telomeric repeats and to associate with telomeres.[24] This again would back the idea of telomere lengthening by recombination. Nonetheless, many questions concerning ALT are still elusive and unsettled.

ALT is the exception in standard telomere maintenance, but it points out that there is certainly not a single solution for all cancers. Also, ALT is not restricted to one type of cancer, so there will always remain a chance that any cancer cell may find its way to circumvent the Hayflick limit.

IV Life Extension Science

i. Molecular Insights

So far, cells that are not subject to the Hayflick limit have been shown to pose a threat to humans' well-being. The general public does for the most part not know that modern medicine, genetics and many beneficial tools of molecular biology were not realized today without immortal cells. Discoveries that led to these advancements were only possible through work with immortal cell lines and go back to the very first cells that were ever cultivated in culture.[1]

The history of cell culturing begins in 1951 with the cervical cancer of a woman called Henrietta Lacks. She was treated at the Johns Hopkins Hospital in Baltimore and during her treatments, tissue samples of her cancer as well as healthy tissue was removed and given to Dr. George Otto Gey. Gey experimented with all kinds of cell types trying to cultivate cells in vitro. Though all his trials had failed before, Henrietta Lacks' cancer cells, now named HeLa, survived and flourished in culture. Gey donated HeLa cells freely to his colleagues all over the world including instructions on how to cultivate them, just for the benefit of science.[3] This provided countless opportunities for new experiments to scientists of various fields and led to HeLa being the most commonly used human cell line up to this day.[25]

HeLa was a lucky strike for science and many developments arose from HeLa research. One of the great advantages of HeLa cells is the possibility to do trials with human cells without having to put human beings in jeopardy of unknown substances, toxins or drugs. HeLa cells are cancer cells, but they still possess most of the functions of a normal human cell. Jonas E. Salk used HeLa cells to test and develop a safe, new polio vaccine. His tests went well and after more trials with real persons his vaccine was soon launched and helped extinguish polio in Western countries. Unsurprisingly, HeLa has also been used to study cancer, too.[3] It was used to link HPV with cervical cancer and even in 2014 researchers still use HeLa cells to study cancer cell behavior, e.g. their metabolic pathways.[26]

Cultivated immortal cells also allowed extensive research into the insides of cells. Gene mapping was developed with HeLa and geneticists discovered that humans have 46 chromosomes instead of 48 chromosomes which was widely believed until then.[3] The first cell to be cloned was a HeLa cell, as well as the first human cell to be fused with an animal cell forming a chimera.[3] Whether all these experiments are desirable and useful is surely arguable. Nonetheless, humankind has greatly benefited from research with HeLa and other immortal cell lines. We have a better understanding of the functioning of cells and therewith we gained better understanding of the functioning of all living organisms including ourselves.

ii. Telomerase Inhibitors

Similar to many other things in life, telomerase is not plain "evil", because it enables cells to overcome the Hayflick limit and allows tumor growth. In contrary, scientists try to find a cure for cancer by targeting telomerase.[27]

The biopharmaceutical company Geron is currently conducting phase 2 clinical trials with their proprietary telomerase inhibitor named imetelstat. It is a short 13-mer oligonucleotide binding sequence specific to the RNA component of human telomerase, hTR. The drug is administered through intravenous infusions; therefore, the oligonucleotide is merged with a lipid group to further membrane permeability.[28] The trials with imetelstat concentrate on patients with myeloproliferative neoplasms (MPNs), a selection made due to preclinical studies that showed a good response to imetelstat in bone marrow hemopoietic cells, hair follicle cells, malignant plasma cells and peripheral blood mononuclear cells.[29] The chosen diseases are all forms of cancer affecting the bone marrow and are attended by massive proliferation of hemopoietic cells; these overproduced cells might also be abnormal.[28]

According to Geron, phase 1 trials evoked remarkable resonance in the participating patients. 89% achieved a complete or partial response rate.[28] Telomerase activity was measurably lower, hemopoietic cell count decreased and administering frequencies of imetelstat could be reduced continuously in all patients.[29] On first sight, these are very promising result.

On the other hand, Geron provides no detailed report about side effects; the company only states on its website that all "Adverse events were manageable and reversible"[28] and no patients left the trial due to side effects. However, since this is a completely new approach and Geron's trial the first of its kind, there are no long-term studies about the effects of telomerase inhibitors in the long run. It is conceivable that patients treated with such inhibitors might suffer from diseases related to short telomeres later on in life.

Another approach to telomerase inhibition has been investigated by Elizabeth Blackburn and her colleagues at the University of California in San Francisco.[15] They introduced mutant-template telomerase to prostate and breast cancer cell lines. The mutant enzymes were made to add wrong tandem repeats to the existing telomere strand. The wrong telomeric sequences restrain the telomere from reforming its t-loops and proteins complex, thus leaving the telomere region unprotected. It will soon be recognized as DNA damage or cause further chromosome aberrations through recombination and trigger DNA damage response leading to apoptosis. As intended, the experiments resulted in decreased cellular viability and significantly increased apoptosis rates. Substantially, low quantities of mutant telomerase were enough to accomplish the intended results and the wild type telomerase did not have to be inhibited.[29] Still, these model in vitro experiments are far from application in real life. One of the biggest issues will be to refine the specificity of telomerase altering drugs, to ensure exact inhibition. Nevertheless, these trials raise strong hope that telomerase might truly be a future target for new cancer therapies.

iii. Anti-Aging Industry

The news that the key to aging might be found sparked a lot of hype about telomeres and telomerase in the anti-aging sector. Today there are several approaches of commercial companies to use telomeres for their products and services.

The first realized telomere service was the measurement of telomere length. The Spanish corporation Life Length was founded in 2010 and since offers determination of biological age by means of telomere length on the individual level as well as for clinical analyses.[30] They use a proprietary Quantitative Fluorescence In Situ Hybridization (Q-FISH) method to estimate the customers biological age based on the percentage of extremely short telomeres. The customer has to provide a tissue or blood sample. Life Length examines interphase nuclei and hybridizes the telomeres with a complementary fluorescent probe. Each probe recognizes a set number of telomeric repeats. Then, the fluorescent emission of every chromosome end is measured. The emission rate is proportional to the hybridized probes and therefore proportional to telomere length. Life Length compares the results to average telomere lengths in their data base and assesses the percentage of very short telomeres for their estimation of the customer's biological age.[31]

Another company named TeloMe Inc. pursues a similar approach. The company uses saliva samples instead of blood or tissue and measures the overall average telomere length.[33] The average telomere length of the customer is then compared to the average telomere length of people in the same age group. The company stopped offering their service to the general public though and confines its services to research studies.[32]

The necessity of this test is debatable. Critics argue that measuring telomere length and foreseeing the future health of someone are not equal. Also, there are no studies that prove a causal connection between telomere length and longevity; they associate them at the most. Another aspect to consider is the possible consequences drawn from the results of a telomere measuring. If the telomeres are average or in even better shape, the person can go on living like before. If the telomeres are found to be very short, there is not much to do about it either.

The solution advised by the named companies is a healthy life style, exercising, little stress and enough sleep.[31,32] It is not surprising and new that a healthy life style enhances chances of a long and healthy life.[33] Nobody needs their telomeres to be measured to learn whether they are living healthy or not, too. Everybody can decide to start living healthier to increase their prospects of a long and healthful life.

Especially oxidative stress is associated with considerable, accelerated telomere attrition.[34] To counter oxidants, some advertise antioxidant supplements and foods as being a natural anti-aging aid. The effectiveness of this thesis is controversial though and lacks thorough scientific proof.[35] A more interesting means to extend life would be a way to elongate telomeres beyond the general "live-healthy" mantra.

A plant called astragalus has aroused attention in natural anti-aging. The roots of astragalus mongholicus have been used in traditional Chinese medicine for 2,000 years to support the immune system and are said to be a natural telomerase activator.[36] The US-American National Institutes of Health's National Center for Complementary and Alternative Medicine (NCCAM) speaks out on astragalus as follows:

"The evidence for using astragalus for any health condition is limited. High-quality clinical trials (studies in people) are generally lacking. There is some preliminary evidence to suggest that astragalus, either alone or in combination with other herbs, may have potential benefits for the immune system, heart, and liver, and as an adjunctive therapy for cancer."[37]

There are some fans of astragalus who believe in its rejuvenating powers.[38] The corporation Telomerase Activation Sciences (TA Sciences) took a further step and sells a pill called TA-65. It is sold as a dietary supplement containing a proprietary, concentrated herbal extract of astragalus. TA-65 is supposed to lengthen telomeres and advertisements promote strengthening of the immune system, younger skin, improved hair quality, an overall better health and a longer life.[39] Since the company sells its product as a nutritional supplement and not as a drug, they are not evaluated by the US Food and Drug Administration (FDA). The company sells its pill successfully all over the world with an annual revenue of US$ 6 million in the United States alone.[40]

The credibility of TA Sciences has incisively been challenged by a former employee who filed a class-action lawsuit against the company accusing it of deceptive practices. Brian Egan was Vice President for Global Sales with TA Sciences. According to his statements in a discrimination complaint he filed in September 2011, he was expected to take TA-65 to tell his clients that he was taking the product, too, and that it was safe and effective. However, Egan was diagnosed with prostate cancer in September 2011. After telling his boss about his diagnosis, he claims to have been immediately fired, because this incident could have ruined the company's reputation and was offered a financial settlement to keep silent. TA Sciences denies this account and contends that Egan already had cancer when he started working with them and that he was fired for meager sales. The conflict was intensified, because Egan told a potential partner of TA Sciences in Spain that he had developed cancer while taking TA-65. The company sued him in return for defamation and said he had lost them US$ 2 million in sales.[40] As far as the media covered this case, the legal battle between the two parties was still going on in 2013. Even though there is no medical evidence that links Egan's diagnosis with his consumption of TA-65, this case casts a shadow over TA Sciences claims. The lack of independent studies concerning the effects of this nutrition supplement adds to the concerns raised against it.[41]

Telomerase activators are a risky endeavor. Assuming, they are effective enough to increase telomere length, they also break down tumor suppressor barriers. Whether TA-65 has this power is in doubt, but there is just one health and there is no reason to take such a foreseeable risk.

V Conclusion

No doubt, cells that are not subject to the Hayflick limit are an invaluable boon to science. Nevertheless, they are the bane of people's lives that fight with cancer.

The abilities to divide indefinitely and to avoid apoptosis are widely held to be vital for tumor growth. This increases the probability that the Hayflick limit is not only the depletion of a cell's natural capacity to divide due to the end replication problem, but also an essential tumor suppressor mechanism.

This mechanism, which is confined by telomere length, usually leads to age-related degenerative diseases, because cells enter senescence, stop dividing, eventually die and the tissue cannot renew itself. From the perspective of evolution, there is no need in nature to support an organism beyond sexual maturity, but in our modern, man-controlled world with high hygiene standards, advanced medicine and no natural enemies, people grow old and are increasingly affected by these diseases.

The enzyme telomerase enables gametes and some other cells with high turnover to proliferate continuously without reaching the Hayflick limit by means of lengthening their telomeres. This ensures that a zygote can grow into a new, healthy body without the risks of extremely short telomeres. The perils of such include cancer, thus telomerase can be seen as an indirect tumor suppressor, too, in these circumstances.

On the other hand, reactivation of telomerase allows most cancer cells to escape their natural growth arrests and to start proliferating uncontrollably, whereas telomerase is not expressed in most somatic cells. Only few cancers use alternative mechanisms of telomere lengthening. This makes telomerase a promising target for cancer therapy. Tremendous advancements have been made to this day and the first telomerase inhibitor has reached the second phase of clinical trials. However, there is still much progress to be made. So far, the telomerase inhibitor can only be used to improve cancers of the bone marrow. Specificity of such drugs is a major issue that will have to be improved in the future. It would need a vector to transport the drug to reach specific malignant cancer cells and to target only those cells that need it. Affecting healthy cells could cause unintentional acceleration of telomere erosion or other unforeseeable adverse events. Also, findings relating to the long-term effects of the telomerase inhibitor in current clinical trials are still awaited.

An enzyme that allows cells to regenerate and proliferate like in early stages of life is of course of great interest for the anti-aging industry. But regarding anti-aging, it should not be forgotten that telomerase is not and cannot be the ultimate remedy for aging, because there is more to ageing than short telomeres. Still, the company TA Sciences sells a concentrated extract of the root astragalus as a dietary supplement that is said to be a natural telomerase activator. The practice has been criticized for it might increase risks of cancer, because it allows for continued cell proliferation, possibly of pre-malignant tumors. The company argues that their product is rather weak, though in turn, one may question why one should take the product if it is not effective enough to promote a considerable change. For those who do believe in its powers and wish to take TA Sciences product it comes down to a trade-off: is the risk of developing cancer matching the possible benefit of an increased health and life span? However, considering these aspects it should never be forgotten that essentially, playing with telomeres is playing with life and death.

Comparing boon and bane of immortalized cells is difficult and cannot lead to a certain result. The role of telomerase in immortalization is a two-edged sword: it can protect the cell from fatal diseases, but it can also establish an opportunity for the very same. Tampering with it is dangerous and should be responsibly pondered on. This is a good analogy for the character of biologically immortal cells in general, too, which have beneficial traits as well as disastrous traits. There cannot be a balance of advantages and disadvantages, because in either case human lives and well-being are at stake. The weighing up is rather an ethical question than a matter of biological pros and cons, because our world is not black and white, but there are always shades.

VI. References

[1] Shay, Jerry W., and Woodring E. Wright. "Hayflick, His Limit, and Cellular Ageing." Nature Oct. 2000. 72-76. Web. 01 Oct. 2013. <http://67-20-95-176.bluehost.com/Hayflick.NatureNotImmortal.pdf>.

[2] Weismann, August. Trans. Arthur E Shipley. "The Duration of Life, 1881." Essays upon Heredity and Kindered Biological Problems. Trans. Arthur E Shipley. Ed. Edward B Poulton and Selmar Schönland. Oxford: Clarendon Press, 1889.

[3] Skloot, Rebecca. The Immortal Life of Henrietta Lacks. New York: Broadway Paperbacks, 2010. Print.

[4] De Magalhães, João P. "The Hayflick Limit." Senescence.info. N.p., 1997. Web. 19 Oct. 2013. <http://www.senescence.info/cell_aging.html>.

[5] Ganten, Detlev, Klaus Ruckpaul, and Antonio Ruiz-Torres. Molekularmedizinische Grundlagen Von Altersspezifischen Krankheiten. Berlin, Heidelberg: Springer Verlag, 2004. Print.

[6] Grundmann, Ekkehard. Das Ist Krebs. München: W. Zuckschwerdt Verlag GmbH, 2007. Print.

[7] Campbell, Neil A., and Jane B. Reece. "Die Molekularen Grundlagen Der Vererbung." Biologie. München: Pearson Studium, 2009. 409-434. Print.

[8] Pecorino, Lauren. Molecular Biology of Cancer. Oxford, UK: Oxford University Press, 2012. Print.

[9] Read, Andrew, and Dian Donnai. New Clinical Genetics. Banbury, UK: Scion Publishing Ltd, 2011. Print.

[10] Bronchud, Miguel H., Mary Ann Foote, William P. Peters, and Murray O. Robinson, eds. Principles of Molecular Oncology. Totowa, New Jersey: Humana Press, 2000. Print.

[11] Xing, Huawei, Dan Liu, and Zhou Songyang. "The Telosome/Shelterin Complex and Its Functions." Genome Biology 9.9 (2008): N. pag. Web. 16 Nov. 2013. <http://www.ncbi.nlm.nih.gov/pmc/articles/PMC2592706/pdf/gb-2008-9-9-232.pdf>.

[12] Cao, Kan, Cecilia D. Blair, Dina A. Faddah, Julia E. Kieckhaefer, Michelle Olive, Michael R. Erdos, Elizabeth G. Nabel, and Francis S. Collins. "Progerin and Telomere Dysfunction Collaborate to Trigger Cellular Senescence in Normal Human Fibroblasts." The Journal of Clinical Investigation 121.7 (2011): 2833-2844. Print.

[13] Halvorsen, Tanya L., Gil Leibowitz, and Fred Levine. "Telomerase Activity Is Sufficient to Allow Transformed Cells to Escape from Crisis." N.p., American Society For Microbiology, 23 Nov. 1998. Web. 29 Oct. 2013. <http://mcb.asm.org/content/19/3/1864.full>.

[14] "The Nobel Prize in Physiology or Medicine 2009". Nobelprize.org. Nobel Media AB 2013. Web. 28 Dec 2013. <http://www.nobelprize.org/nobel_prizes/medicine/laureates/2009/>

[15] Nihvcast. "Telomerase and the Consequences of Telomere Dysfunction". Youtube. Youtube, LLC. 07 Dec 2010. Web. 14 Sep 2013. <http://www.youtube.com/watch?v=fleIXeAFIO0>.

[16] Meyerson, Matthew. "Role of Telomerase in Normal and Cancer Cells." Journal Of Clinical Oncology. American Society Of Clinical Oncology, 1 July 2000. Web. 29 Nov. 2013. <http://jco.ascopubs.org/content/18/13/2626.full>.

[17] Tarsounas, Magdalena. "It's Getting HOT at Telomeres." The EMBO Journal 32.12 (2013): 1655-1657. Print.

[18] Wynford-Thomas, David, and David Kipling. "Cancer and the Knockout Mouse." Nature 389(1997): 551f. Print.

[19] "Video Player". Nobelprize.org. Nobel Media AB 2013. Web. 28 Dec 2013. <http://www.nobelprize.org/mediaplayer/index.php?id=1214>.

[20] Krupp, Guido, and Reza Parwaresch. Telomerases, Telomeres and Cancer. New York City, New York, USA: Kluwer Academic/ Plenum Publishers, 2002. Print.

[21] Reddel, Roger R. "The Role of Senescence and Immortalization in Carcinogenesis." Carcinogenesis 21.3 (2000): 477-484. Print.

[22] Bertorelle, R, M Briarava, E Rampazzo, L Biasini, M Agostini, I Maretto, S Lonardi, M L. Friso, C Mescoli, V Zagonel, D Nitti, A De Rossi, and S Pucciarelli. "Telomerase Is an Independent Prognostic Marker of Overall Survival in Patients with Colorectal Cancer." British Journal of Cancer 108(2013): 278-284. Print.

[23] Durusoy, Mebeccel, and Kamile Öztürk. "Methods Used in Evaluating Telomerase Activity." Tübitak. 2 Feb. 2001. Web. 7 Dec. 2013. <http://journals.tubitak.gov.tr/medical/issues/sag-01-31-5/sag-31-5-1-0102-4.pdf>.

[24] Chung, Inn, Sarah Osterwald, Katharina I. Deeg, and Karsten Rippe. "PML Body Meets Telomere - the Beginning of an ALTernate Ending?" Nucleus 3.3 (2012): 263-275. Web. 7 Dec. 2013. <https://www.landesbioscience.com/journals/nucleus/2012NUCLEUS0003R.pdf>.

[25] Raditsch, Lena. "Havoc in Biology's Most-Used Human Cell Line." European Molecular Biology Laboratory. EMBL Heidelberg, 11 Mar. 2013. Web. 26 Dec. 2013. <https://www.embl.de/aboutus/communication_outreach/media_relations/2013/130311_Heidelberg/>.

[26] Marín-Hernández, Alvaro, Sayra Y. López-Ramírez, Juan C. Gallardo-Pérez, Sara Rodríguez-Enríquez, Rafael Moreno-Sánchez, and Emma Saavedra. Systems Biology of Metabolic and Signaling Networks. Berlin, Heidelberg: Springer Verlag, 2014. Web. 06 Jan. 2014. <http://link.springer.com/chapter/10.1007/978-3-642-38505-6_9#page-2>.

[27] Naylor, RM, DJ Baker, and JM Van Deursen. "Senescent Cells: a Novel Therapeutic Target for Aging and Age-Related Diseases." Nature 93.1 (2013): 105-116. Print.

[28] Geron Corporation. "R & D Imetelstat." Geron. Geron Corporation, n.d. Web. 29 Dec. 2013. <http://www.geron.com/imetelstat>.

[29] Tefferi, Ayalew. "Imetelstat Sodium in Treating Patients with Primary or Secondary Myelofibrosis." ClinicalTrial.gov. Mayo Clinic, 18 Nov. 2012. Web. 7 Dec. 2013. <http://clinicaltrials.gov/ct2/show/study/NCT01731951>.

[30] Hackenbroch, Veronika. "Tests Zur Lebenserwartung: Gute Geschäfte Mit Den "Zündschnüren Des Todes"." Spiegel, 18 May 2011. Web. 24 May 2013. <http://www.spiegel.de/wissenschaft/medizin/tests-zur-lebenserwartung-gute-geschaefte-mit-den-zuendschnueren-des-todes-a-763393.html>.

[31] Life Length. "Life Length for Individuals." Life Length, n.d. Web. 15 Dec. 2013. <http://www.lifelength.com/individuals.html>.

[32] TeloMe Inc.. "Frequently Asked Questions." TeloMe. N.p., 2012. Web. 28 Dec. 2013. <http://www.telome.com/faq.php>.

[33] Marchant, Jo. "Spit Test Offers Guide to Health." Nature. 28 May 2011. Web. 28 Dec. 2013. <http://www.nature.com/news/2011/110528/full/news.2011.330.html>.

[34] Babizhayev, MA, EL Savel'yeva, SN Moskvina, and YE Yegorov. "Telomere Length Is a Biomarker of Cumulative Oxidative Stress, Biologic Age, and an Independent Predictor of Survival and Therapeutic Treatment Requirement Associated with Smoking Behavior." Web. 17 Dec. 2013. <http://www.ncbi.nlm.nih.gov/pubmed/20228673>.

[35] Chun, Ock, Balz Frei, Christopher Gardner, Lee Alekel, and John Killen. "Get the Facts: Antioxidants and Health - an Introduction." National Center For Complementary And Alternative Medicine. U.S. Department Of Human & Health Services, Nov. 2013. Web. 27 Dec. 2013. <http://nccam.nih.gov/health/antioxidants/introduction.htm>.

[36] Vitaminum GbR. "Astragalus Wirkung." Vitaminum GbR, 2012. Web. 23 Dec. 2013.
<http://www.astragalus-membranaceus.com/wirkung.html>.
[37] NCCAM. "Herbs at a Glance - Astragalus." National Center For Complementary And Alternative
Medicine. U.S. Department Of Health & Human Services, Apr. 2012. Web. 23 Dec. 2013.
<http://nccam.nih.gov/health/astragalus>.
[38] Vitaminum GbR. "Verjüngung - Anti-Aging Mit Astragalus." Vitaminum GbR, 2012. Web. 23 Dec.
2013. <http://www.astragalus-membranaceus.com/verjuengung.html>.
[39] T.A. Sciences. Cell Rejuvenation Through Telomerase Reactivation. N.p., 2010. Web. 06 Jan.
2014. <http://www.tasciences.com/>.
[40] Borrell, Brendan. "Lawsuit Challenges Anti-Aging Claims." Nature 2 Aug. 2012: 2. Print.
[41] Andrews, William, and Michael West. "Turning on Immortality: the Debate over Telomerase
Activation." Life Extension Magazine. Life Extension Foundation For Longer Life, Aug. 2009. Web.
06 Jan. 2014. <http://www.lef.org/magazine/mag2009/aug2009_Turning-on-Immortality-The-
Debate-Over-Telomerase-Activation_01.htm>.
[42] Narwaria, Malti, Archana Shrivastav, and B R. Shrivastav. "Mini Review Telomerase - a
Biomarker in the Carcinogenesis and Diagnosis." ARPN Journal of Science and Technology 2.8
(2012): 733-737. Web. 2 Jan. 2014.
<http://www.ejournalofscience.org/archive/vol2no8/vol2no8_7.pdf>.
[43] Gan, Y, J Lu, A Johnson, MG Wientjes, DE Schuller, and JL Au. "A Quantitative Assay of
Telomerase Activity." PubMed. Pharmaceutical Research, 18 Apr. 2001. Web. 7 Dec. 2013.
<http://www.ncbi.nlm.nih.gov/pubmed/11451036>.
[44] Blackburn, Elizabeth H. "Structure and Function of Telomeres." Nature 350(1991): 570-573. Print.
[45] Aubert, Geraldine, and Peter M. Lansdorp. "Telomeres and Aging." Physiol Rev 350(2008): 557-
579. Print.

VII. Bibliography

Andrews, William, and Michael West. "Turning on Immortality: the Debate over
Telomerase Activation." Life Extension Magazine. Life Extension Foundation For Longer
Life, Aug. 2009. Web. 06 Jan. 2014.
<http://www.lef.org/magazine/mag2009/aug2009_Turning-on-Immortality-The-Debate-
Over-Telomerase-Activation_01.htm>.

Aubert, Geraldine, and Peter M. Lansdorp. "Telomeres and Aging." Physiol Rev
350(2008): 557-579. Print.

Babizhayev, MA, EL Savel'yeva, SN Moskvina, and YE Yegorov. "Telomere Length Is a
Biomarker of Cumulative Oxidative Stress, Biologic Age, and an Independent Predictor of
Survival and Therapeutic Treatment Requirement Associated with Smoking Behavior."
Web. 17 Dec. 2013. <http://www.ncbi.nlm.nih.gov/pubmed/20228673>.

Bertorelle, R, M Briarava, E Rampazzo, L Biasini, M Agostini, I Maretto, S Lonardi, M L.
Friso, C Mescoli, V Zagonel, D Nitti, A De Rossi, and S Pucciarelli. "Telomerase Is an
Independent Prognostic Marker of Overall Survival in Patients with Colorectal Cancer."
British Journal of Cancer 108(2013): 278-284. Print.

Blackburn, Elizabeth H. "Structure and Function of Telomeres." Nature 350(1991): 570-573. Print.

Borrell, Brendan. "Lawsuit Challenges Anti-Aging Claims." Nature 2 Aug. 2012: 2. Print.

Bronchud, Miguel H., Mary Ann Foote, William P. Peters, and Murray O. Robinson, eds. Principles of Molecular Oncology. Totowa, New Jersey: Humana Press, 2000. Print.

Campbell, Neil A., and Jane B. Reece. "Die Molekularen Grundlagen Der Vererbung." Biologie. München: Pearson Studium, 2009. 409-434. Print.

Cao, Kan, Cecilia D. Blair, Dina A. Faddah, Julia E. Kieckhaefer, Michelle Olive, Michael R. Erdos, Elizabeth G. Nabel, and Francis S. Collins. "Progerin and Telomere Dysfunction Collaborate to Trigger Cellular Senescence in Normal Human Fibroblasts." The Journal of Clinical Investigation 121.7 (2011): 2833-2844. Print.

Chun, Ock, Balz Frei, Christopher Gardner, Lee Alekel, and John Killen. "Get the Facts: Antioxidants and Health - an Introduction." National Center For Complementary And Alternative Medicine. U.S. Department Of Human & Health Services, Nov. 2013. Web. 27 Dec. 2013. <http://nccam.nih.gov/health/antioxidants/introduction.htm>.

Chung, Inn, Sarah Osterwald, Katharina I. Deeg, and Karsten Rippe. "PML Body Meets Telomere - the Beginning of an ALTernate Ending?" Nucleus 3.3 (2012): 263-275. Web. 7 Dec. 2013. <https://www.landesbioscience.com/journals/nucleus/2012NUCLEUS0003R.pdf>.

De Magalhães, João P. "The Hayflick Limit." Senescence.info. N.p., 1997. Web. 19 Oct. 2013. <http://www.senescence.info/cell_aging.html>.

Durusoy, Mebeccel, and Kamile Öztürk. "Methods Used in Evaluating Telomerase Activity." Tübitak. 2 Feb. 2001. Web. 7 Dec. 2013. <http://journals.tubitak.gov.tr/medical/issues/sag-01-31-5/sag-31-5-1-0102-4.pdf>.

Gan, Y, J Lu, A Johnson, MG Wientjes, DE Schuller, and JL Au. "A Quantitative Assay of Telomerase Activity." PubMed. Pharmaceutical Research, 18 Apr. 2001. Web. 7 Dec. 2013. <http://www.ncbi.nlm.nih.gov/pubmed/11451036>.

Ganten, Detlev, Klaus Ruckpaul, and Antonio Ruiz-Torres. Molekularmedizinische Grundlagen Von Altersspezifischen Krankheiten. Berlin, Heidelberg: Springer Verlag, 2004. Print.

Geron Corporation. "R & D Imetelstat." Geron. Geron Corporation, n.d. Web. 29 Dec. 2013. <http://www.geron.com/imetelstat>.

Grundmann, Ekkehard. Das Ist Krebs. München: W. Zuckschwerdt Verlag GmbH, 2007. Print.

Hackenbroch, Veronika. "Tests Zur Lebenserwartung: Gute Geschäfte Mit Den "Zündschnüren Des Todes"." Spiegel, 18 May 2011. Web. 24 May 2013. <http://www.spiegel.de/wissenschaft/medizin/tests-zur-lebenserwartung-gute-geschaefte-mit-den-zuendschnueren-des-todes-a-763393.html>.

Halvorsen, Tanya L., Gil Leibowitz, and Fred Levine. "Telomerase Activity Is Sufficient to Allow Transformed Cells to Escape from Crisis." N.p., American Society For Microbiology, 23 Nov. 1998. Web. 29 Oct. 2013. <http://mcb.asm.org/content/19/3/1864.full>.

Life Length. "Life Length for Individuals." Life Length, n.d. Web. 15 Dec. 2013. <http://www.lifelength.com/individuals.html>.

Kim, Moses M., Melissa A. Rivera, Inna L. Botchkina, Refaat Shalaby, Ann D. Thor, and Elizabeth H. Blackburn. "A Low Threshold Level of Expression of Mutant-template Telomerase RNA Inhibits Human Tumor Cell Proliferation." PNAS 98.14 (2001): 7982-7987. Print.

Krupp, Guido, and Reza Parwaresch. Telomerases, Telomeres and Cancer. New York City, New York, USA: Kluwer Academic/ Plenum Publishers, 2002. Print.

Marchant, Jo. "Spit Test Offers Guide to Health." Nature. 28 May 2011. Web. 28 Dec. 2013. <http://www.nature.com/news/2011/110528/full/news.2011.330.html>.

Marín-Hernández, Alvaro, Sayra Y. López-Ramírez, Juan C. Gallardo-Pérez, Sara Rodríguez-Enríquez, Rafael Moreno-Sánchez, and Emma Saavedra. Systems Biology of Metabolic and Signaling Networks. Berlin, Heidelberg: Springer Verlag, 2014. Web. 06 Jan. 2014. <http://link.springer.com/chapter/10.1007/978-3-642-38505-6_9#page-2>.

Meyerson, Matthew. "Role of Telomerase in Normal and Cancer Cells." Journal Of Clinical Oncology. American Society Of Clinical Oncology, 1 July 2000. Web. 29 Nov. 2013. <http://jco.ascopubs.org/content/18/13/2626.full>.

Narwaria, Malti, Archana Shrivastav, and B R. Shrivastav. "Mini Review Telomerase - a Biomarker in the Carcinogenesis and Diagnosis." ARPN Journal of Science and Technology 2.8 (2012): 733-737. Web. 2 Jan. 2014. <http://www.ejournalofscience.org/archive/vol2no8/vol2no8_7.pdf>.

Naylor, RM, DJ Baker, and JM Van Deursen. "Senescent Cells: a Novel Therapeutic Target for Aging and Age-Related Diseases." Nature 93.1 (2013): 105-116. Print.

NCCAM. "Herbs at a Glance - Astragalus." National Center For Complementary And Alternative Medicine. U.S. Department Of Health & Human Services, Apr. 2012. Web. 23 Dec. 2013. <http://nccam.nih.gov/health/astragalus>.

NCSU Libraries. "Citation Builder." NC State University. N.p., n.d. Web. 26 May 2013. <http://www.lib.ncsu.edu/citationbuilder/cite.php?source=website>.

Nihvcast. "Telomerase and the Consequences of Telomere Dysfunction". Youtube. Youtube, LLC. 07 Dec 2010. Web. 14 Sep 2013. <http://www.youtube.com/watch?v=fIeIXeAFIO0>.

Oshima, Alice, and Ann Hogue. Writing Academic English. White Plains, NY: Pearson Longman, 2006. Print.

Pecorino, Lauren. Molecular Biology of Cancer. Oxford, UK: Oxford University Press, 2012. Print.

Raditsch, Lena. "Havoc in Biology's Most-Used Human Cell Line." European Molecular Biology Laboratory. EMBL Heidelberg, 11 Mar. 2013. Web. 26 Dec. 2013. <https://www.embl.de/aboutus/communication_outreach/media_relations/2013/130311_He idelberg/>.

Read, Andrew, and Dian Donnai. New Clinical Genetics. Banbury, UK: Scion Publishing Ltd, 2011. Print.

Reddel, Roger R. "The Role of Senescence and Immortalization in Carcinogenesis." Carcinogenesis 21.3 (2000): 477-484. Print.

Shay, Jerry W., and Woodring E. Wright. "Hayflick, His Limit, and Cellular Ageing." Nature Oct. 2000. 72-76. Web. 01 Oct. 2013. <http://67-20-95-176.bluehost.com/Hayflick.NatureNotImmortal.pdf>.

Skloot, Rebecca. The Immortal Life of Henrietta Lacks. New York: Broadway Paperbacks, 2010. Print.

T.A. Sciences. Cell Rejuvenation Through Telomerase Reactivation. N.p., 2010. Web. 06 Jan. 2014. <http://www.tasciences.com/>.

Tarsounas, Magdalena. "It's Getting HOT at Telomeres." The EMBO Journal 32.12 (2013): 1655-1657. Print.

Tefferi, Ayalew. "Imetelstat Sodium in Treating Patients with Primary or Secondary Myelofibrosis." ClinicalTrial.gov. Mayo Clinic, 18 Nov. 2012. Web. 7 Dec. 2013. <http://clinicaltrials.gov/ct2/show/study/NCT01731951>.

TeloMe Inc.. "Frequently Asked Questions." TeloMe. N.p., 2012. Web. 28 Dec. 2013. <http://www.telome.com/faq.php>.

"The Nobel Prize in Physiology or Medicine 2009". Nobelprize.org. Nobel Media AB 2013. Web. 28 Dec 2013. <http://www.nobelprize.org/nobel_prizes/medicine/laureates/2009/>.

"Video Player". Nobelprize.org. Nobel Media AB 2013. Web. 28 Dec 2013. <http://www.nobelprize.org/mediaplayer/index.php?id=1214>.

Vitaminum GbR. "Astragalus Wirkung." Vitaminum GbR, 2012. Web. 23 Dec. 2013. <http://www.astragalus-membranaceus.com/wirkung.html>.

Vitaminum GbR. "Verjüngung - Anti-Aging Mit Astragalus." Vitaminum GbR, 2012. Web. 23 Dec. 2013. <http://www.astragalus-membranaceus.com/verjuengung.html>.

Weismann, August. Trans. Arthur E Shipley. "The Duration of Life, 1881." Essays upon Heredity and Kindered Biological Problems. Trans. Arthur E Shipley. Ed. Edward B Poulton and Selmar Schönland. Oxford, UK: Clarendon Press, 1889. 5-35. Web. 02 Oct. 2013. <http://archive.org/stream/essaysuponhered02shipgoog#page/n3/mode/2up>.

Wynford-Thomas, David, and David Kipling. "Cancer and the Knockout Mouse." Nature 389(1997): 551f. Print.

Xing, Huawei, Dan Liu, and Zhou Songyang. "The Telosome/Shelterin Complex and Its Functions." Genome Biology 9.9 (2008): N. pag. Web. 16 Nov. 2013. <http://www.ncbi.nlm.nih.gov/pmc/articles/PMC2592706/pdf/gb-2008-9-9-232.pdf>.